EUREKA!
THE BIOGRAPHY OF AN IDEA

BICYCLE

BY LORI HASKINS HOURAN • ILLUSTRATED BY AARON CUSHLEY

KANEPRESS

AN IMPRINT OF BOYDS MILLS & KANE

New York

For Michael, a whiz on two wheels —LHH

To Cian and Fionn. Get on yer bike —AC

Special thanks to Andrew Ritchie, editor of *The Boneshaker*,
the journal of the Veteran-Cycle Club

Kane Press
An imprint of Boyds Mills & Kane, a division
of Astra Publishing House
kanepress.com
Printed in China

Library of Congress Cataloging-in-Publication Data
Names: Houran, Lori Haskins, author. | Cushley, Aaron, illustrator.
Title: Bicycle / by Lori Haskins Houran ; illustrated by Aaron Cushley.
Description: First edition. | New York : Kane Press, an imprint of Boyds Mills & Kane,
[2021] | Series: Eureka! | Audience: Ages 4-8 | Summary: "A nonfiction 'biography'
of the bicycle, an everyday object that has become ubiquitous, starting with its possible
origins after the eruption of Mount Tambora and up through modern bicycles to suit
every rider"—Provided by publisher.
Identifiers: LCCN 2020054052 (print) | LCCN 2020054053 (ebook) | ISBN 9781635923933 (hc)
| ISBN 9781635923940 (pbk) | ISBN 9781635924749 (ebook)
Subjects: LCSH: Bicycles—History—Juvenile literature.
Classification: LCC TL412 .H68 2021 (print) | LCC TL412 (ebook) | DDC 629.227/209—dc23
LC record available at https://lccn.loc.gov/2020054052
LC ebook record available at https://lccn.loc.gov/2020054053

10 9 8 7 6 5 4 3 2 1

WHO LIKES BIKES?

Just about everybody! People ride bicycles for exercise, to get around town . . . and for fun!

Bicycles have been around for over two hundred years. An inventor named Karl Drais made the first bike in 1817. What gave Drais the idea? Some say it was a volcano.

Yes, a volcano.

INDONESIA, 1815

Mount Tambora wasn't just any volcano. When it blew, it BLEW. Ash filled the sky and spread around the globe. The ash blocked part of the sun's light . . . and kept blocking it for a year.

Newspapers called 1816 "The Year Without a Summer." In America, snow fell in July. In Europe, crops stopped growing. Oats got so scarce, people had to give up their horses. It cost too much to feed them.

Without horses, people had to walk everywhere. Walking was awfully slow.

There had to be a faster way to get around! Karl Drais, a German inventor, set out to make one.

Drais built a wooden vehicle with two wheels, one in front of the other.

It did not have pedals. Instead, Drais powered it by pushing off the ground with his feet. He called it a running machine.

GERMANY, 1818

Drais's invention worked pretty well. It went about eight miles an hour—more than twice as fast as walking. But people were afraid to try it. They were sure they'd fall right over!

Drais explained again and again that the machine was easy to ride. It just took a little practice.

Not many folks listened.

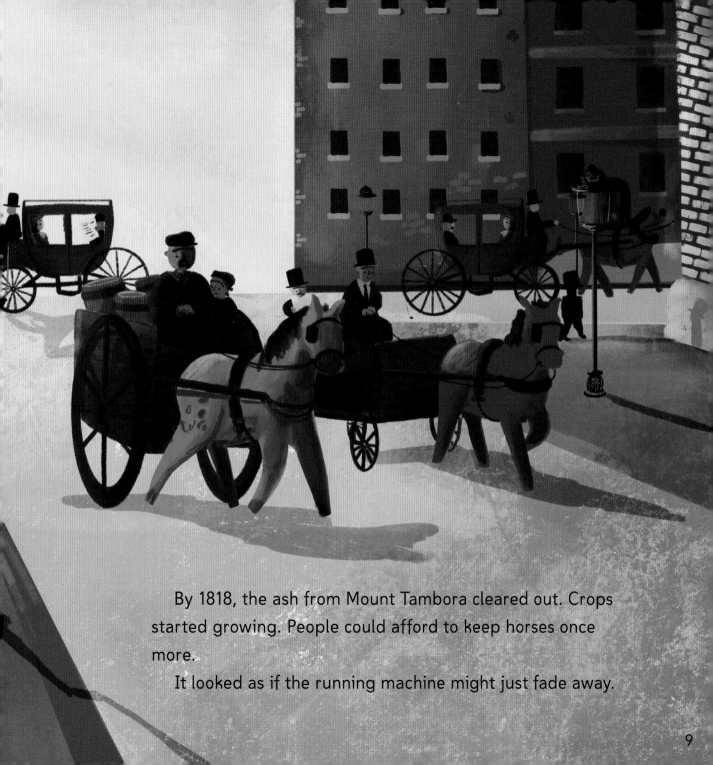

By 1818, the ash from Mount Tambora cleared out. Crops started growing. People could afford to keep horses once more.

It looked as if the running machine might just fade away.

But not everyone forgot about Drais's machine.

Young riders in England and France had taken a fancy to it. The English called it a "dandy horse." French riders named it the *velocipede*, from the Latin words for "quick" and "feet."

Someone in France, possibly a blacksmith named Pierre Michaux, got a clever idea. He added pedals to the front wheel of the velocipede.

Now the velocipede was faster—and more fun!—than before.

ENGLAND, 1868

The new velocipede caught on. First in Europe, then across the sea in America.

Along the way, it got a funny nickname—the boneshaker.

Why? Because the velocipede wasn't comfortable. AT ALL.
Its wheels were made of hard wood and even harder iron.
Riders felt every bump in the road.

And, oh, there were bumps!

AMERICA, 1870

Other inventors tinkered with the idea. Three-wheeled tricycles became popular—and not just for kids!

Then came the high wheeler. Its tall front wheel made it a bit smoother to ride than the boneshaker.

The high wheeler had a different flaw, though. It was tippy. Terribly tippy. And when riders fell off, they had a *long* way to fall.

In America, so many people got hurt that a new word was coined. A "header" meant a hard fall off a high wheeler . . . headfirst!

Most people who rode high wheelers were daredevils. Who else would risk it?

J. K. Starley, an English inventor, wanted to make a bike for everyone. So he created the Rover safety bicycle.

The front and back wheels of the Rover were roughly the same size. That meant the bike wasn't tippy.

The pedals were in the middle of the bike, not on the front wheel. That made the Rover easy to pedal.

Suddenly, riding a bicycle wasn't so risky. It was safe and simple.

••• THE WHEEL DEAL •••

What happens when you ride your bike? Your foot pushes the **pedal**. The pedal turns a **crank** in the middle of the bike with a **chain** wrapped around it. That chain starts to move, faster and faster.

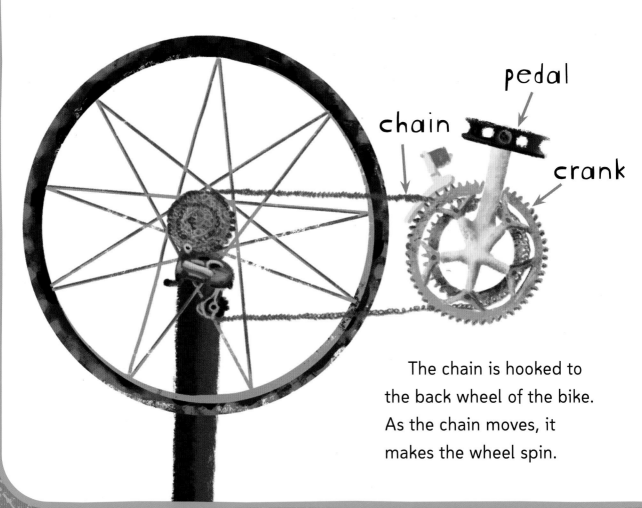

chain

pedal

crank

The chain is hooked to the back wheel of the bike. As the chain moves, it makes the wheel spin.

Every time the wheel spins, you roll ahead. With a single push of the pedal, you can zoom forward about five feet. That's a lot farther than you'd go with a single step!

IRELAND, 1888

The bicycle had come a long way. Still, it had one problem. Well, two problems—the wheels.

By now, they were made of solid rubber. That was better than the old wood and iron wheels. But not much.

John Boyd Dunlop watched his young son rattle down the street on a tricycle. Could he give the boy a gentler ride?

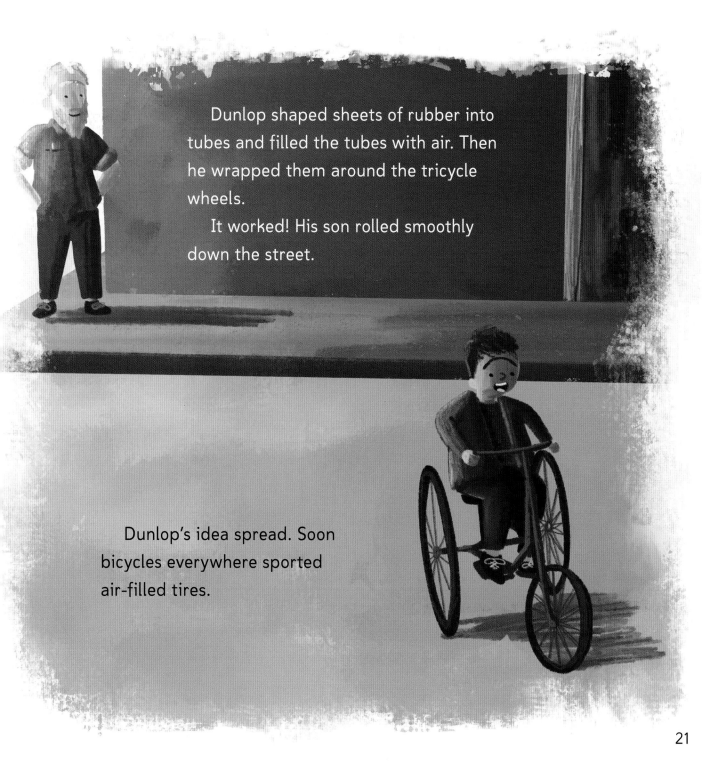

Dunlop shaped sheets of rubber into tubes and filled the tubes with air. Then he wrapped them around the tricycle wheels.

It worked! His son rolled smoothly down the street.

Dunlop's idea spread. Soon bicycles everywhere sported air-filled tires.

AMERICA, 1895

At long last, bikes were safe, easy to pedal, *and* comfortable.
More people started to ride. Old folks. Young folks.
Gentlemen. Ladies.

Companies built special bicycles just for women. The bikes dipped low, so ladies could climb on with their long skirts.

A few women said, *No, thanks*. They put on pants instead. That raised some eyebrows!

CHINA, 1950

By 1940, lots of adults in America had moved on to something faster than bikes—cars.

But in many countries, bicycles were still the best way to get around.

In 1950, a Chinese worker named Huo Baoji designed a simple black bike called the Flying Pigeon. Nearly every household in China bought one.

The Flying Pigeon became the most popular vehicle in the world!

The streets of China are still filled with Flying Pigeons. More than 500 million of them have been made.

Today, bikes come in countless colors and styles—stunt bikes, mountain bikes, racing bikes, even bikes that people pedal with their hands. There's a bicycle for just about everybody!

It's hard to imagine life without bicycles. And to think, it all may have started . . .

. . . with a volcano.

••• BIKE FACTS •••

You'll fly By!!!
Built to be FASTER!
TRY the NEW ROVER Safety Bicycle!!!

• The word *bicycle* was first used in the 1860s. "Bi" means two, for the two wheels on a bike.

• Amsterdam has more bikes than people!

• Before the Wright brothers flew airplanes, they ran a bicycle shop. They built their first plane in the shop.

• The Tour de France is the world's most famous bicycle race. It takes 23 days, and racers ride more than 2,000 miles.

••• PEDAL POWER •••

The bicycle is the most energy-efficient form
of transportation on Earth. Its engine is your body.
Its fuel is the food you eat.

That apple you ate at breakfast has around 100 calories. If
you walked, those calories would give you enough energy to walk
about one mile. But if you biked, those calories could carry you up
to five miles!